2019 Alachua County Hurricane Dorian After Action Report / Improvement Plan

Alachua County NFARC/ARES(R) Voice Net Practice Session

Gordon L. Gibby KX4Z

and the volunteers of the Alachua County ARES(R) group, and

North Florida Amateur Radio Club

Copyright © 2019   Gordon L. Gibby MD KX4Z

Approved for publication by the assembled members of the
North Florida Amateur Radio Club
on Sept 11 2019.

All rights reserved, except the work may be copied with attribution to improve and enhance amateur radio emergency communications preparedness.

ISBN: 9781692586805

DEDICATION

This After Action/Improvement Plan document is dedicated to all the volunteers of Alachua County who have worked so hard to provide a better volunteer backup response.

Wendell Wright putting in new shelving for our new Radio Room

CONTENTS

Chapter	Title	Page
	Acknowledgments	
1	Introduction	1
2	Alachua County Merged Timeline	3
3	Observations, Recommendations, Alachua County Improvement Plan	11
	APPENDIX: Alachua County ARES (R) Boiler-Plate Incident Action Plan	15
	About the North Florida Amateur Radio Club	25

ACKNOWLEDGMENTS

The North Florida Amateur Radio Club would like to acknowledge all the work done by the Alachua County Emergency Management Group to further backup amateur radio support capabilities, and in particular the incredible support provided by Col. Huckstep, W4JIR, of the Alachua County Sheriff.

Alachua County Emergency Management has graciously facilitated our amateur radio license courses and EC-001 training with the use of their building. We especially note the outstanding assistance of Dalton Herding.

Alachua County 2019 Hurricane Dorian AAR/IP

1 INTRODUCTION

Hurricane Dorian caused major death and destruction in the Caribbean, and significantly threatened Florida. Floridians remembered our recent hurricane experience, and stocked up a bit more in advance this time. Gas still became scarce for about 2 days in Alachua County. The University shut down, expecting the arrival on Tuesday -- and then the track of the Hurricane unexpectedly slowed, and changed, and Florida was largely spared.

One cannot "turn off" an incident -- considerably different from an Exercise -- and everyone in Alachua ARES (R) had to make preparations for the possibility of another significant event. This gave us a chance to test our previous improvements, as you'll see in this report, and think through some systems that we hadn't really understood well before.

Gordon L. Gibby and Alachua County Volunteers

2 ALACHUA COUNTY MERGED TIMELINE

MONDAY AUGUST 26	
Monday August 26	Evening advisories from national weather service on earliest reasonable arrival of tropical winds: 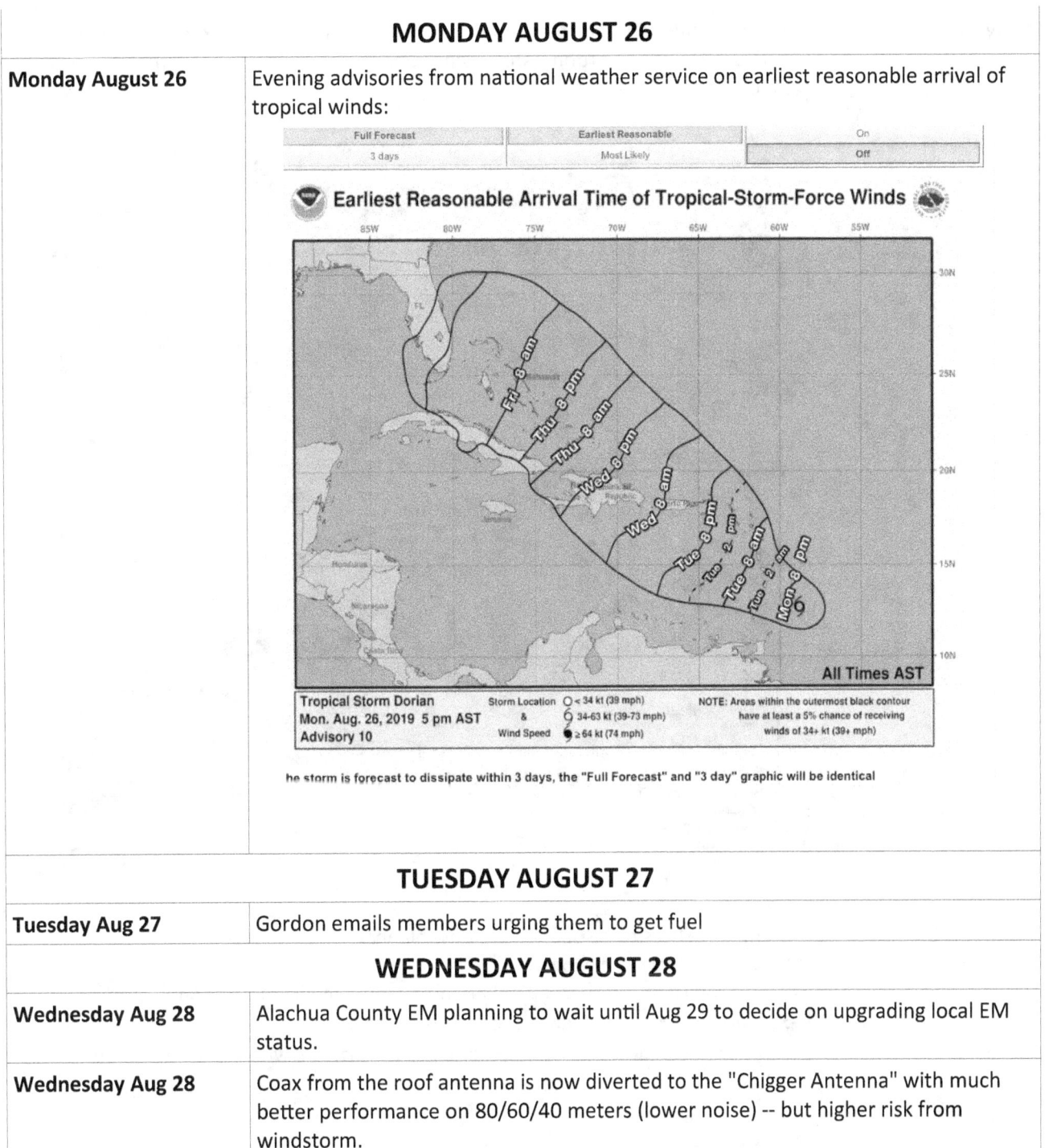
TUESDAY AUGUST 27	
Tuesday Aug 27	Gordon emails members urging them to get fuel
WEDNESDAY AUGUST 28	
Wednesday Aug 28	Alachua County EM planning to wait until Aug 29 to decide on upgrading local EM status.
Wednesday Aug 28	Coax from the roof antenna is now diverted to the "Chigger Antenna" with much better performance on 80/60/40 meters (lower noise) -- but higher risk from windstorm.

Wednesday Aug 28	Gainesville SARNET connection still not working

THURSDAY AUGUST 29

Thursday Aug 29	Florida Gov. declares statewide emergency -- Alachua County waiting until Friday Aug 30 to decide on opening shelters. EOC meeting Alachua County at 1 PM; local ARES (R) net meets at 8PM (Leland NCS)
Thursday Aug 29	Gordon forwards SHARES info to Leland

FRIDAY AUGUST 30

Friday Aug 30	University of Florida cancels classes for Tuesday Sept 3 due to expected arrival of Hurricane Dorian. Gordon is discussing getting coax from our roof antenna down as a backup.
Friday Aug 30	Potential shelters to be opened are now planned: Order of shelters to be opened: #1 will be Easton, #2 MLK (Waldo road) both of which are pet friendly #3 Senior center, able to take people with medical issues; #4 Westwood middle school if Senior Center overflows with medically needy people. Other Issues:

RADIOS
Only one go box is completely populated (the prototype).
800 MHz police radios WITH POWER SUPPLY will be individually positioned at shelters that don't get the only complete go box.
Col. Huckstep was upset that others were not done, but I reassured him that there is very little chance they would be needed as we had AT LEAST 3 vhf/uhf go-boxes in RIDGID boxes at the EOC and only *1* was used when we have many, many shelters open.

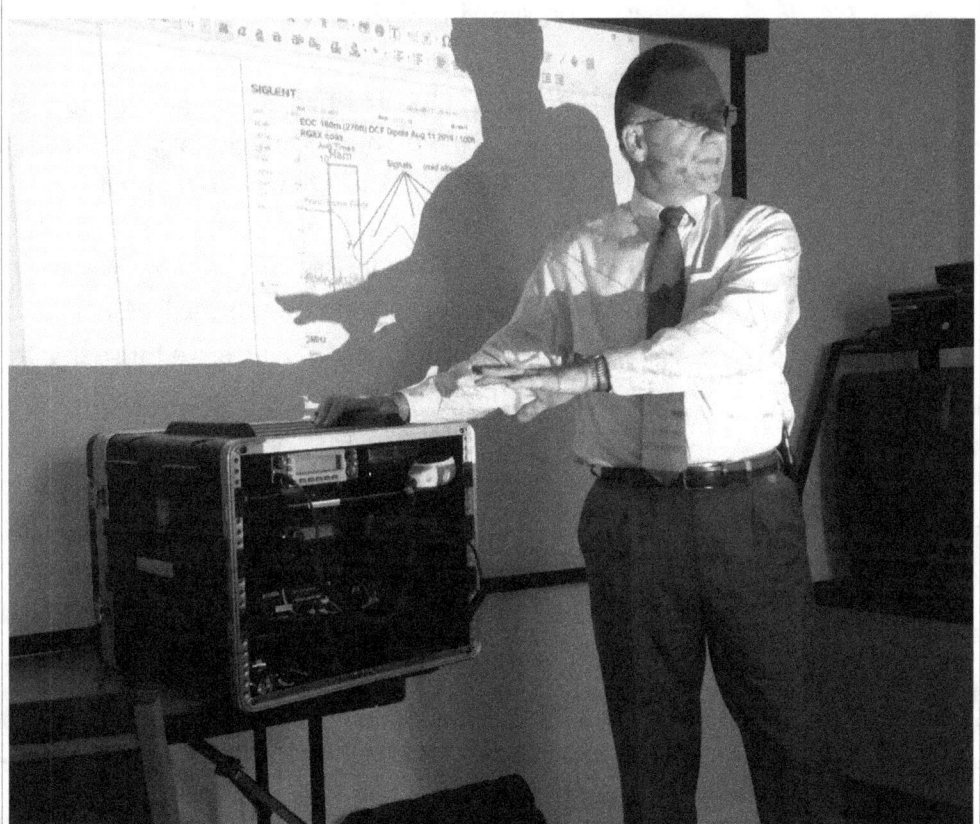

Col. Huckstep presenting the go-box systems developed by Alachua County employees.

BATTERIES: I would suggest that we disperse those precious batteries to the shelters as they open in case power drops. The police radios appear to have bare wires. Hams would need to simply connect to the 12V battery if that becomes an issue.

COAX: hams will need to bring coax (PL-259 on both ends) to go from ham radios to the VHF/UHF SO-239 on the wall to the already installed exterior

	antenna.

**Police radios have N connectors. Col. Huckstep is going to look into getting coax with N connectors to go from the police radio to the wall connection....that hadn't been quite recognized (he thinks)
Anyone crazy enough like me to run HF needs to bring their own HF antenna and coax to reach the feed-thru and as we learned, a ladder to reach the other side. Obviously not many are going to need to do that so I would NOT worry about it at all.

Decision on WHAT / WHEN to open is likely to come MONDAY.

BADGED VOLUNTEERS: the badges are not yet out. Every shelter will have a list of the approved volunteers and you will go there and just show your identification and they will look you up on their list.

If you don't have a suitable VHF/UHF radio -- you will need to get one of the RIDGID boxes we made, which are at the EOC and fill out a ICS-211e line and sign it please. I will ask Leland to arrange for such a form to appear down there and be "find-able" for you to sign out radios.

Our plan is to drop a cable from the rooftop antenna on Monday at 11 AM |
| **Friday Aug 30** | Jeff Capehart adds a forwarding rule for RACES email so he will get copies, and subsequently arranges for other local leadership to get copies as well
Alachua county has not yet made any decisions on closing any County facilities. |
| **SATURDAY AUGUST 31** ||
| **Saturday August 31** | Gordon drafts a nearly-complete ICS-201 for activation of the ARES (R) group -- with "TBD" left in all the spots for shelters, times, and volunteer assignments.

This form (an ICS-205 boilerplate specifically designated for the ARES (R) group alone) was shown to the Alachua County Chief Deputy and to the Alachua County Emergency Manager and both thought it was excellent. A copy was then sent to the Alachua County EM for his records at his request.

Due to meeting schedules and availability, antenna work to add a cable from the rooftop antenna gets moved up to 0930 Sat Morning. Accomplished with a cable hanging down to a bush by the back door. The rope tensions on the roof antenna are adjusted properly. Col. Huckstep, Ryan Lee, Gordon Gibby, Leland Gallup.

- |

Coaxial cable routed to the roof top antenna.

SUNDAY SEPT 1	
Sunday Sept 1	NFL Section ARES (R) advances to Level III 10 AM
Sunday Sept 1	Request made by KX4Z to the JS8 group on groups.io, to have volunteers join @ALACHUA for testing of relay possibilities.
MONDAY SEPT 2	
Monday Sept 2	P4 waiver provided by FCC Gordon works out in theory a way to provide a forwarding system to get to the Ocala SARNET since the Gainesville system isn't working Alachua county is now "out of the cone" so efforts are stepping down a bit.

Monday Sept 2
1330LOC
	JS8 Testing at the time of day that was WORST for North Florida SSB practice exercise:
	
	• 80 meter JS8: successfully captured message back from K4MVR, cached a message for Leland
	• 40 meter JS8: two responses to @ALACHUA from volunteers;
	  ○ W4CAT (Tennessee) SNR -1
	  ○ KN4CRD (Georgia) SNR -8
	• 30 meter JS 8 -- no responses.

	Gordon drove south on 441 until he found a location where one could just reach the Ocala access to the SARNET -- and still communicate by VHF to Alachua County. The location was marked by a windmill at the top of a large hill. Even a 5 watt walkie talkie to a mag mount on the top of the truck would generally reach the Ocala system for getting into the SARNET from this location.

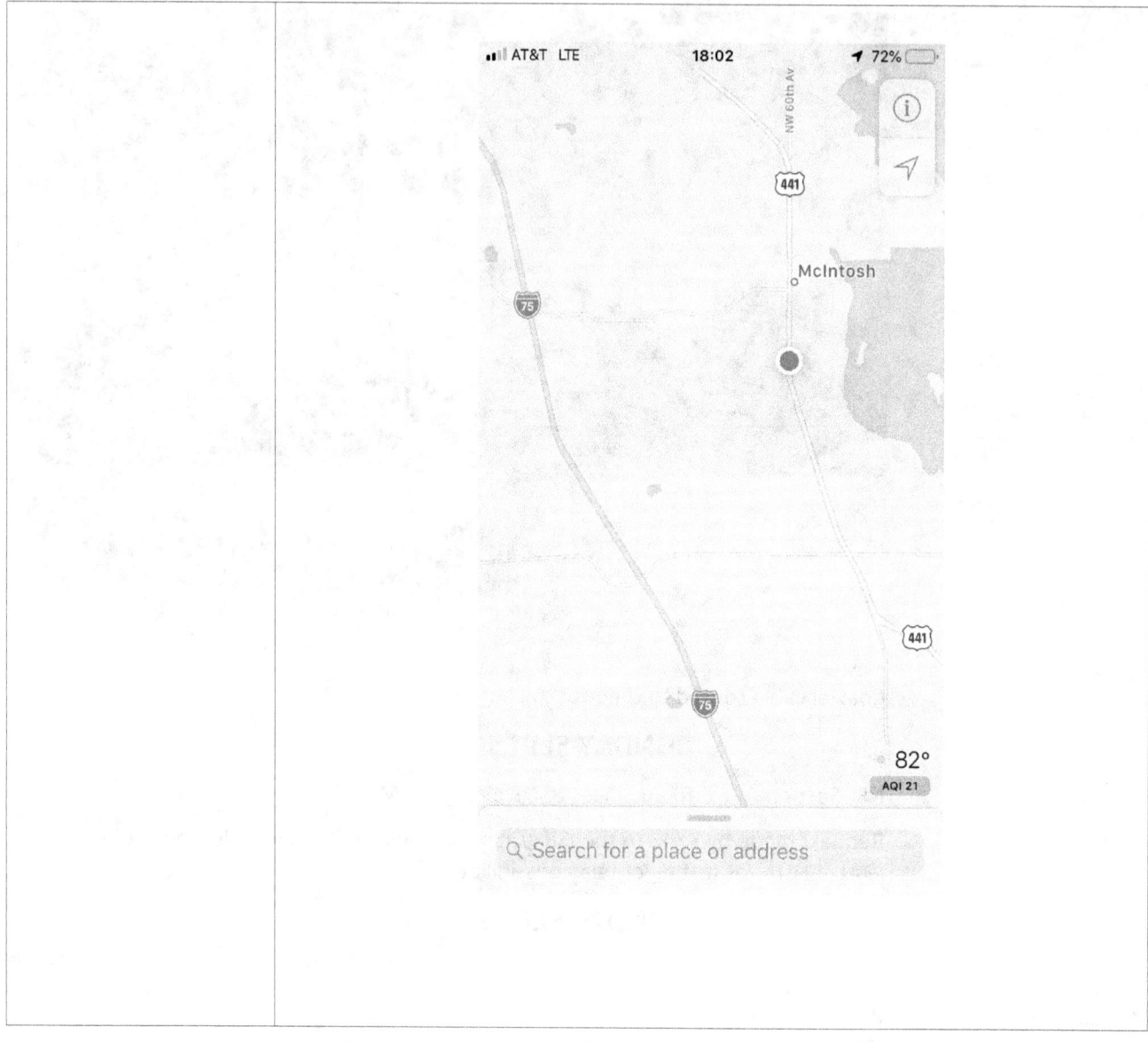

	 There are likely many ways to leverage this structure at that hill top, in an emergency.
Monday Sept 2 7:30 PM	**SUCCESS at sending radio gram cached over JS8.** Message cached at K4MVR and picked up by AA3YB --- "NR 3 KX4Z 6 Newberry FL....." and so on. FIRST RADIOGRAM BY OUR GROUP OVER JS8

Monday Sept 2 -- time uncertain, likely 7:30 PM	• 80 meters -- K4MVR reports SNR -22
	• 40 meters
	   ◦ KC1GU (Massachusetts) reports SNR -2
	   ◦ K4KDR (Virginia) SNR 1
	   ◦ W1FVB (New Hampshire) SNR +0
	
	• 30 meters
	   ◦ K8CJM (Ohio) SNR -14
	   ◦ W1FVB (New Hampshire) -13
	   ◦ W7SUA (Arizona) SNR -6

TUESDAY SEPT 3	
Tue Sept 3 6AM	JS8 40 meters W7SUA (Arizona)

Tue Sept 3	NFL Section ARES (R) advances to level II at 10 AM
WEDNESDAY SEPT 4	
Wed. Sept 4	NFL Section ARES (R) decreases to Level III at 8 PM
THURSDAY SEPT 5	
Thur Sept 5	NFL Section ARES (R) stand down 10 AM

Alachua ARES (R) - Standby only (Level 3)

No shelters activated, but 4 pre-determined.

(MLK, Newberry/Archery, Sr. Center, Westwood MS)

We had 6 available ARES (R) members on Monday; 5 Tuesday; 3 Wed. Most had concerns over whether their employer was closing or remain open, or had other obligations that made them pending or conditional for future availability after the holiday. If we had been requested to open 4 shelters, we would have needed 2 people at the EOC, and 1-2 people at each shelter. That would have been 6 minimum; 10 preferred.

EOC staffed only by ESF-5. Only 7am-7pm shift. Gordon/Leland visited EOC. SitReps emailed to staff.

"Near 0 percent chance of impact".

Highest wind was 40mph Tue @ 2:20pm near UF.

An ICS-205 document was sent ahead of time.

Karl sent around a consolidated Florida SITREP.

FLORIDA ARES (R) Net was Active 0800-1800 daily in 2 hour net control slots.

SARNET was open until 8pm 9/4 with NCS in 4 hour slots.

3 OBSERVATIONS, RECOMMENDATIONS, & ALACHUA COUNTY IMPROVEMENT PLANS

Note: The Alachua County NFARC group can only make improvement plans for its own local efforts.

Observation	Recommendations for amateur radio operators	Improvement Plan (for NFARC / Alachua County ARES (R))
1. Alachua County RACES volunteers still do not have any formal badges, though important decision on which volunteers are approved were made, a huge step forward.		Continue to encourage Alachua county EM to complete the badging process.
2. **Excellent progress on a lower-noise "chigger antenna" in the woods -- but everyone recognizes increased risk to windstorm, therefore the coax from the roof top antenna was added-- excellent progress. But a formal 2nd coax and switch are needed.**		Plans are tentative at this point, need to be firmed up for the final coax. We anticipate this will conclude our antenna redevelopment that involves county efforts. We may make further improvement to the antenna in the woods, but that should not require County effort
3. Alachua County RACES is using borrowed HF transceiver for user-friendly operations, is limited to 100 watts output by transceiver, loaned antenna tuner, and limitations of the antenna.		Aggressive improvement plan is forming: 1. Continue to use loaner transceiver until better option available. 2. A chassis of an SB-200 is becoming available for rebuild and modification for combined ham/shares operation. Pursue rebuilding this amplifier to allow for 400-600 watts output SSB. 3 Pursue a higher-power antenna tuner 4. Materials have already arrived to allow a higher-power Balun to be constructed to improve the

		power handling capabilities of the antenna. 5. Consider adding coax traps to the antenna to provide for 4-5 MHz operation., for Federal interoperability
4. SARNET continues to be a problem for Alachua County because the Gainesville DOT equipment's receiver system is not functioning. An experiment carried out prior to the hurricane demonstrated that a 35 watt signal within 100 yards of the antenna still did not create full quieting. That had been known by the GARS Repeater committee for a year, but their requests were not fulfilled. There are political issues in the way of getting this fixed.		A plan was worked out for a relay computer/patch system to allow the EOC to connect to the Ocala SARNET -- but a by-the-book interpretation of the SARNET rules disallows this solution, and requires a live human to provide relaying. A plan and location were worked out to handle this exigency, but it would be much better if the Gainesville system worked. **UPDATE: As of Sept 11 2019-- the Gainesville SARNET system is now repaired.**
5. SARNET is at risk of jamming. A malicious interferer who would not identify was encountered during Alachua County testing to try and reach Ocala.	Amateur radio operators created quite the response when DOT indicated they did not wish to service the SARNET. It would be wise for Section Officials to stay on top of this and keep bringing it back to the hams when things like our repeater don't work for a year or more.	We just cannot put all our hopes on the SARNET Having HF, and digital and a multitude of skills remain important. That is why getting the EOC backup radio systems up to speed have received such a priority for ARES (R).

APPENDIX:

BOILER PLATE ALACHUA ARES (R)-SPECIFIC INCIDENT ACTION PLAN

This was boiler plate content created to make it much easier to get ALL the information out to our people in the event that we had to begin shelter operations. The ICS system was designed expressly for this purpose, as you learn in ICS-300 etc. Alachua County never opened any shelters in this incident, so this document was never completed/sent out for real -- but the Emergency Manager was very pleased and excited to see our level of preparedness and professionalism.

Gordon L. Gibby & Alachua County Volunteers

INCIDENT BRIEFING (ICS 201) ALACHUA ARES (R) ONLY

1. Incident Name: Hurricane Dorian ARES (R)	2. Incident Number: 2019-1	3. Date/Time Initiated: Date: 8/31/19 Time: Noon

4. Map/Sketch (include sketch, showing the total area of operations, the incident site/area, impacted and threatened areas, overflight results, trajectories, impacted shorelines, or other graphics depicting situational status and resource assignment):

Easton Newberry Archery- 24880 NW 16th Ave, Newberry, FL 32669 (352) 472-2388

MLK Multi Purpose (@Citizens Field) 1028 NE 14th St, Gainesville, FL 32601 Phone:(352) 334-5053

Senior Center: 5701 NW 34th St, Gainesville, FL 32653 Phone:(352) 265-9040

Westwood Middle School 3215 NW 15th Ave, Gainesville, FL 32605   (352) 955-6718

5. Situation Summary and Health and Safety Briefing (for briefings or transfer of command): Recognize potential incident Health and Safety Hazards and develop necessary measures (remove hazard, provide personal protective equipment, warn people of the hazard) to protect responders from those hazards.

SHELTER OPENING TIME	TBD
EARLIEST TROP WINDS	MON 8AM
LIKELIEST TROP WINDS	TUE 8 AM

6. Prepared by: Name: G. Gibby_____ Position/Title: Vol_____ Signature:_____

ICS 201, Page 1   Date/Time: Saturday 0930 LOC_____

Alachua County 2019 Hurricane Dorian AAR/IP

INCIDENT BRIEFING (ICS 201)

1. Incident Name: Hurricane Dorian ARES (R)	2. Incident Number:	3. Date/Time Initiated: Date: Time:

7. Current and Planned Objectives:

A. Plans for staffing 4 shelters

B. Expansion plans for up to 2 additional shelters

C. Plans for staffing EOC

D. Have everyone practice a situation report on Saturday

E. Assigned volunteers LOCATE their shelter

F. Assigned volunteers GET READY

G. DISBURSE BATTERIES AND NEEDED RADIOS with ICS-211e RECORDS

8. Current and Planned Actions, Strategies, and Tactics: (ALL TIMES ARE LOCAL)

Time:	Actions:
1100	Contact and Notify assigned volunteers
1800	Disburse needed go-boxes and batteries to volunteers with signed ICS-211e Form -- arranged with each volunteer individually
Today	Volunteers to **locate** their shelter -- **NOTIFY ALACHUA-EC@WINLINK.ORG OR CALL JEFF CAPEHART WHEN TASK COMPLETED**
Today	Volunteers to put together their gear for visiting shelter upon opening
Today	All ARES (R) members to send in at least ONE SHARES SPOT-REP situation reports for practice -- USING EITHER WINLINK OR VOICE a) if by WINLINK: to ALACHUA-DATA@WINLINK.ORG and ALACHUA-EC@WINLINK.ORG b) if by voice (radiogram equivalent) send to W4UFL
Today	All ARES (R) members to PRINT and/or SAVE this document, as future versions may show only CHANGES to it to allow delivery over radio.
Today	As Available, 146.820 monitored
	REMEMBER THAT WEAPONS ARE NOT ALLOWED INSIDE SHELTERS

6. Prepared by: Name: _____ Position/Title: _____ Signature: _____

ICS 201, Page 2   Date/Time: _____

Gordon L. Gibby & Alachua County Volunteers

INCIDENT BRIEFING (ICS 201)

1. Incident Name:	2. Incident Number:	3. Date/Time Initiated: Date: Time:

9. Current Organization (fill in additional organization as appropriate)

Alachua County Emergency Manager Hal Grieb

HAM RADIO ARES (R)

Alachua County Emergency Coordinator Jeff Capehart W4UFL (352) 219 3901

Alachua County Assistant Emergency Coordinators (Participating)

Leland Gallup AA3YB (443) 538-0314

Susan Halbert KG4VWI (352) 495 1931

Gordon Gibby KX4Z Cell (352) 246 6183

Allan West W4JD (352) 328 2359

Corrections? Send them to docvacuumtubes@gmail.com

6. Prepared by: Name: _____ Position/Title: _____ Signature: _____
ICS 201, Page 3   Date/Time: _____

Alachua County 2019 Hurricane Dorian AAR/IP

INCIDENT BRIEFING (ICS 201)

1. Incident Name: Hurricane Dorian ARES (R)	2. Incident Number:	3. Date/Time Initiated: Date: 8/31 Time: Noon

10. Resource Summary:

Resource	Resource Identifier	Date/Time Ordered	ETA	Arrived	Notes (location/assignment/status)
Charged AGM Batteries	BATT			☐	make arrangements with Gordon / Leland to pick up a battery on Sat
VHF/UHF go boxes	VHF			☐	if needed, make arrangesments with Gordon/Leland to pick up on Sat
Coax cable to reach SO-239	COAX			☐	**You must provide coax to reach the wall mounted SO239 at your shelter**
FLASHLIGHT	LIGHT			☐	Please have a backup illumination
SHARES spot rep	SPOTREP			☐	Please locate this message template under SHARES
800 MHZ RADIO	800MHz			☐	These will be delivered individually to each shelter, hopefully with a cable. N connector output to N connector on wall. if you have adapters....bring them
FOOD, ETC	FOOD			☐	Please take with you what you want for food, beyond what might be provided at shelter
				☐	
				☐	
				☐	
				☐	
				☐	
				☐	

6. Prepared by: Name: _____ Position/Title: _____ Signature: _____

ICS 201, Page 4   Date/Time: _____

Updated by FDA 2/2011

HERE IS HOW TO FIND THE SHARES SPOT REPORT:

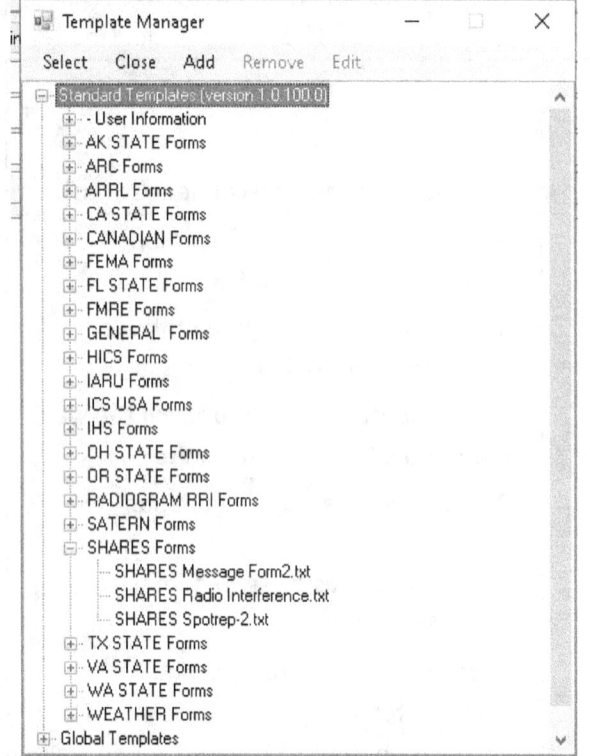

WE ARE GOING TO ASK EVERY ARES (R) MEMBER TO SEND AT LEAST ONE AND PREFERABLY TWO OF THESE SPOT REPORTS TO

(TACTICAL CALL SIGN)

ON SATURDAY FOR PRACTICE.

IN THE ADDITIONAL COMMENTS SECTION PUT

TOTAL SHELTER OCCUPANTS:

AND MAKE UP A NUMBER.

SUGGESTED VOICE SPOT REPORT FORMAT

If you need to file a spot report via voice, here is a suggested format:

\multicolumn{9}{c}{Amateur Radio "Radiogram"}
NR 1	PRECED R	HX	Stn of Origin K4AAA	Check 30	Place of Origin GAINESVILLE FL		Time Filed	Date Filed AUG 31

Addressed TO:

__JEFF CAPEHART W4UFL _____

Message Received At:
Station: _____   Phone: _____
Name/Addr: _____
_____

email_____
phone_____
<BT>
CITY GAINESVILLE X LANDLINE YES
X CELL YES X BROADCAST
YES X TV STATIONS YES
X PUBLIC WATER YES X
COMMERCIAL POWER YES X INTERNET
YES X SIX PERSONS HERE

<BT>
SIGNATURE: _____JOHN DOE K4AAA_____

RCVD FROM	DATE	TIME	SENT TO	DATE	TIME

ASSIGNMENT LIST (ICS 204)

1. Incident Name: Hurricane Dorian ARES (R)	2. Operational Period: Date From: Date To: Time From: Time To:	3. Branch: Division: Group:

4. Operations Personnel:	Name	Contact Number(s)
Operations Section Chief:		
Branch Director:		
Division/Group Supervisor:		

5. Resources Assigned:

Resource Identifier	Leader	Persons # of	Contact (e.g., phone, pager, radio frequency, etc.)	Reporting Location, Special Equipment and Supplies, Remarks, Notes, Information
			TBD	EASTON NEWB
			TBD	SENIOR CENTER
			TBD	MLK CENTER
			TBD	WESTWOOD MID
			Leland Gallup AA3YB	EOC
			Susan Halbert KG4VWI	NF4AC
	VARIABLE		ALACHUA-DATA@WINLINK.ORG	ALACHUA-DATA
	VARIABLE		ALACHUA-EC@WINLINK.ORG	ALACHUA-EC
	Current net control		ALACHUA-NCS@WINLINK.ORG	ALACHUA-NCS
			TBD	Additional Shelter
			TBD	Additional Shelter

6. Work Assignments:
Saturday: FIND your location. Assemble your gear. Plan your arrival.
SEND 1 or 2 spot-reports

7. Special Instructions:

8. Communications (radio and/or phone contact numbers needed for this assignment):
Name / Function   Primary Contact: indicate cell, pager, or radio (frequency/system/channel)
_____ / _____   _____
_____ / _____   _____

9. Prepared by: Name: _____	Position/Title: _____	Signature: _____
ICS 204	IAP Page ____	Date/Time: _____

Alachua County 2019 Hurricane Dorian AAR/IP

INCIDENT RADIO COMMUNICATIONS PLAN   ICS-205

INCIDENT RADIO COMMUNICATIONS PLAN (ICS-205)

1. Incident Name: Hurricane Dorian ARES (R)

2. Date/Time Prepared: Date: / Time:

3. Operational Period Date From: / Date To: / Time From: / Time To:

4. Basic Radio Channel Use:

Zone Grp.	Ch #	Function	Channel Name / Trunked Radio System Talkgroup	Assignment	RX Freq. N or W	RX TONE / NAC	TX FREQ N or W	TX TONE / NAC	MODE (A, D, or M)	Remarks
		Tactical	K4GNV82	Ham	146.82	NA	146.22	123	A	Primary / Com Net
		Tactical	K4GNV68	Ham	146.685	NA	146.085	123	A	Secondary Net
		Tactical	SIMPLX 49	Ham	146.49	CSQ	146.49	CSQ	A	Simplex local
		Logistic	GNV PKT07	Ham	145.07	NA	145.07	NA	D	VHF PACKET-GNV KX4Z-10 via NEWB
		Tact	HF VOICE	Ham	3.950 LSB	NA	3.950 LSB	NA	A	North Florida Ares (R)Net
		Neigh	FRS2GMRS	Any	462.5875	NA	462.5875	NA	A	HAM NEIGHBOR WATCH ch 2
		Logist	OCALAPKT	Ham	145.03	NA	145.03	NA	D	Digi to KX4Z-10 via W4DFU-8
		Tact	W4DFU-91	Ham	146.91	CSQ	146.91	CSQ	A	Tertiary repeater
		Tact	Nat 2m Call	Ham	146.52	CSQ	146.52	CSQ	A	2 meter calling freq
		Log	WINLINK	Ham	TBD		TBD		D	Use software applic.

5. Special Instructions

6. Prepared By (Communications Unit Leader) Name Gordon Gibby   Signature

ICS 205   IAP PAGE   Date / Time 0600 8/31

GLG 5/2018

ICS-205A

1. Incident Name:	2. DATE / TIME PREPARED: Date: Time:	3. OPERATIONAL PERIOD Date From: Time From: Date TO: Time TO:
3. Basic Local Communications Information:	NOTE: TBD = TO BE DETERMINED	
Incident Assigned Position	Name (Alphabetized)	Method(s) of Contact (phone, pager, cell, etc.)
EASTON NEWBERRY	TBD	TBD
SENIOR CENTER	TBD	TBD
MLK CENTER	TBD	TBD
WESTWOOD MIDDLE	TBD	TBD
4. Prepared by: Name: Position/Title: Signature:		
ICS 205A	**IAP Page** _____	Date/Time: _____

ICS 211A CHECK IN LIST (COMMUNICATIONS)	1. INCIDENT NAME:	2. DATE:		3. INCIDENT NUMBER:	4. CHECK IN LOCATION	
5. INFORMATION							
NAME	CALL SIGN	AGENCY or LOCATION	TIME IN	TIME OUT	HOURS	CONTACT INFORMATION	REMARKS
ICS 211A Alachua	6. NUMBER OF PAGES: _____ of _____	7. PREPARED BY (RESOURCE UNIT):		8. MISSION NUMBER	

THIS FORM IS PROVIDED FOR YOUR USE IN ANY TIMEKEEPING

ABOUT THE NORTH FLORIDA AMATEUR RADIO CLUB

The North Florida Amateur Radio Club was formed in order to better support the ARES(R) mission in Alachua County. The formation of the club allowed acquisition of a club callsign and also liability insurance -- two crucial assets when carrying out simulated exercises on public or private property with permission, and when using WINLINK radio email (which requires callsigns for its own email addresses).

The club maintains a web site (https://www.qsl.net/nf4rc/) and is very active in carrying out NIMS-compliant exercises and writing them up afterwards, in HSEEP format when possible. These are published on the club website and usually also on Amazon as soft-cover books.

Previous publications resulting from the activities of this club include:

2017 Hurricane Exercise
https://qsl.net/nf4rc/2017AlachuaCountyCreateSpaceAfterActionReport.pdf
https://www.amazon.com/Alachua-County-Hurricane-Action-Reports/dp/1548062200

2017 Steinhatchee Storm Exercise
https://qsl.net/nf4rc/2017AlachuaCountyCreateSpaceSteinhatcheeAAR.pdf
https://www.amazon.com/Steinhatchee-Storm-How-Puerto-Rico-volunteer/dp/1978441509

2018 Wacassassa Wildfire
https://www.qsl.net/nf4rc/2018/2018 AlachuaCounty Waccasassa Wildfire Excersize.pdf
https://www.amazon.com/Waccasassa-Wildfire-Exercise-Alachua-Reports/dp/1721727817

2018 Emergency Symposium
https://www.amazon.com/Amateur-Radio-Emergency-Communications-Symposium/dp/1983678805

2019 Emergency Conference
https://www.amazon.com/Amateur-Radio-Emergency-Communications-Conference/dp/1791865941

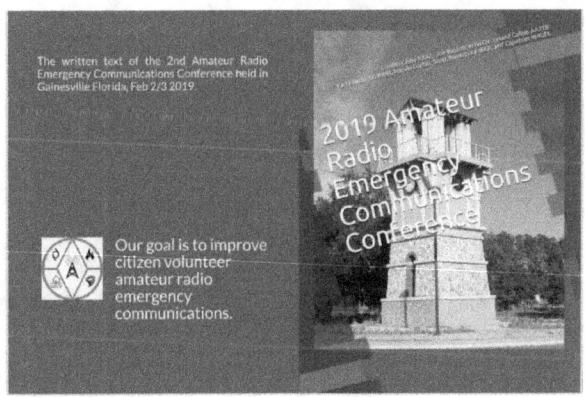

www.ingramcontent.com/pod-product-compliance
Lightning Source LLC
Chambersburg PA
CBHW081708220526
45466CB00009B/2911